ICME-13 Topical Surveys

Series editor

Gabriele Kaiser, Faculty of Education, University of Hamburg, Hamburg, Germany

Katherine Safford-Ramus · Pradeep Kumar Misra
Terry Maguire

The Troika of Adult Learners, Lifelong Learning, and Mathematics

Learning from Research, Current Paradoxes, Tensions and Promotional Strategies

Katherine Safford-Ramus
Saint Peter's University
Jersey City, NJ
USA

Pradeep Kumar Misra
Chaudhary Charan Singh University
Meerut
India

Terry Maguire
National Forum for the Enhancement
  of Teaching and Learning
  in Higher Education
Dublin
Ireland

ISSN 2366-5947   ISSN 2366-5955   (electronic)
ICME-13 Topical Surveys
ISBN 978-3-319-32807-2   ISBN 978-3-319-32808-9   (eBook)
DOI 10.1007/978-3-319-32808-9

Library of Congress Control Number: 2016937347

Printed on acid-free paper

This Springer imprint is published by Springer Nature
The registered company is Springer International Publishing AG Switzerland

Main Topics You Can Find in This "ICME-13 Topical Survey"

- A thought-provoking discussion about adult learners, lifelong learning, and mathematics and their beneficial but challenging relationship;
- An extensive literature review of "adult mathematics education" and presentation of synopsis of the six emerging themes;
- A critical discussion about recent developments in adult mathematics/numeracy in terms of policies, provisions, and challenges;
- A detailed discussion of some of the paradoxes and tensions that are emerging as adult learning mathematics becomes increasingly regulated in a rapidly developing digital world;
- A discussion about five potential strategies to promote lifelong learning of mathematics among adult learners.

Contents

Chapter 1
Introduction

The troika of adult learners, lifelong learning, and mathematics is a unique one. They all are significant in individual capacity and when intermingled makes a useful combination for the benefit of humanity and society. Adult learners are a significant proportion of the world population, lifelong learning is vital to keep one active and engaged, and mathematical learning is important to success in different walks of life. Adult learners practicing lifelong mathematical learning are supposed to be more productive, economically active, and individually satisfied. Instead of these obvious benefits, it is an irony that promotion of lifelong learning of mathematics among adult learners is not high on national and international agenda. In this backdrop, the present book mirrors the troika of adult learners, lifelong learning, and mathematics from three angles. The first angle reveals that adult learners, lifelong learning, and mathematics are significant in individual capacity and when intermingled makes a useful combination for benefit of humanity and society. Adult learners are a significant proportion of world population, lifelong learning is vital to keep one active and engaged, and mathematical learning is important to get success in different walks of life. Adult learners practicing lifelong mathematical learning are supposed to be more productive, economically active, and individually satisfied. The observation ends on the note that instead of the obvious benefits, promotion of lifelong learning of mathematics among adult learners is not high on national and international agenda.

The second angle confirms that the literature base on adults learning mathematics has grown substantially over the past twenty-five years. It is not, however, mainstream and much of the research lies hidden in doctoral dissertations and conference proceedings. Summarization of the results of a literature review and examination of journal articles indexed as "adult mathematics education" present before us six themes related to adult learning mathematics. While, the third angle looks at recent developments in adult mathematics/numeracy in terms of policy and provision and discusses some of the paradoxes and tensions that are emerging as

© The Author(s) 2016
K. Safford-Ramus et al., *The Troika of Adult Learners, Lifelong Learning, and Mathematics*, ICME-13 Topical Surveys, DOI 10.1007/978-3-319-32808-9_1

adult learning mathematics becomes increasingly regulated in a rapidly developing digital world. This observation further lead us towards a number of very useful and pertinent questions like—How can the research domain of adult learning mathematics develop to be able to connect with the emerging disciplines associated with e.g., technology development? How is numeracy conceptualised and what does this mean for adult learners of mathematics and for their teachers? What kinds of adult mathematics provision are being developed? How is this being translated into practice and what provision is needed for developing teacher knowledge, skills and competence? After mirroring the troika from all these angles, the book presents five potential strategies for promotion of lifelong learning of mathematics among adult learners and hope that academicians, researchers and policy makers will take cognizance and find out useful ways, techniques and policies to support adult learning mathematics.

Chapter 2
Survey on State-of-the-Art

2.1 The Troika of Adult Learners, Lifelong Learning, and Mathematics

What are "adult learners"? Adult learners are usually defined as a very diverse group (typically ages 25 and older) with a wide range of abilities, educational and cultural backgrounds, responsibilities and job experiences (Southern Regional Education Board 2015). 'Adult' is interpreted as referring to people who start, resume or continue their education in formal, informal or non-formal settings, beyond the normal age of schooling in their societies (ICME 13 2015). Looking into the future we see a rising number of adult learners. Adult learners are different from traditional college students. Many adult learners have responsibilities (e.g., families and jobs) and situations (e.g., transportation, childcare, domestic violence and the need to earn an income) that can interfere with the learning process. Most adults enter educational programs voluntarily and manage their classes around work and family responsibilities. Additionally, most adult learners are highly motivated and task-oriented (Merriam and Caffarella 1999). Talking about the characteristics of adult learners, Pappas (2013) observes,

> Adults are characterized by maturity, self-confidence, autonomy, solid decision-making, and are generally more practical, multi-tasking, purposeful, self-directed, experienced, and less open-minded and receptive to change. All these traits affect their motivation, as well as their ability to learn.

These adult learners face many challenges in their lives, such as multiple careers, fewer stable social structures to rely on, living longer, and dealing with aging parents. The past is less helpful for them as a guide for living in the present. Their life is complex due to career, family, and other personal choices (Cercone 2008, p. 139). These adult learners are supposed to have some kind of support system to keep them active, productive, and receptive to face the challenges and complexities of life. Learning throughout life is one such support system.

© The Author(s) 2016
K. Safford-Ramus et al., *The Troika of Adult Learners, Lifelong Learning, and Mathematics*, ICME-13 Topical Surveys, DOI 10.1007/978-3-319-32808-9_2

Continuing learning helps one to learn new tricks, adapt well to changing socio-economic conditions, and emerge as a better citizen. The real value of lifelong learning is to enable people to equip themselves to act, to reflect and to respond appropriately to the social, political, economic, cultural and technological challenges they face throughout their lives (Medel-Añonuevo et al. 2001).

2.1.1 Lifelong Learning for Adult Learners: Need and Significance

The concept of lifelong learning stresses that learning and education are related to life as a whole—not just to work—and learning throughout life is a continuum that should run from cradle to grave. According to this concept, lifelong learning refers to all kinds of formal education and training (whether or not they carry certification); and can occur anywhere including education or training institutions, the workplace (on or off the job), the family, or cultural and community settings (Misra 2012, p. 289). Lifelong learning, according to Royce (1999, p. 149), "Aims to give students the skills to go on learning throughout life and also positive attitudes towards learning which accept and even welcome change and new learning." In this sense, lifelong learning supports the development of knowledge and competences to enable each citizen to adapt to the knowledge-based society and actively participate in all spheres of social and economic life. In the European Commission (2001, p. 9), Lifelong Learning (LLL) is defined as

> All learning activity undertaken throughout life, with the aim of improving knowledge, skills and competence, within a personal, civic, social and/or employment-related perspective.

The other definition of Lifelong Learning given by Jarvis (2006, p. 134) is very relevant with reference to adult learners,

> The combination of processes throughout a life time whereby the whole person-body (genetic, physical and biological) and mind (knowledge, skills, attitudes, values, emotions, beliefs and senses) – experiences social situations, the perceived content of which is then transformed cognitively, emotively or practically (or through any combination) and integrated into the individual person's biography resulting in a continually changing (or more experienced) person.

The essence of these definitions helps us to claim that lifelong learning offers different opportunities for adult learners to learn in a variety of contexts—in educational institutions, at work, at home and through leisure activities (Misra 2012). Schuller and Watson (2009) advocates that the right to learn throughout life is a human right and vision about a society in which learning plays its full role in personal growth and emancipation, prosperity, solidarity and local and global responsibility. Therefore, provisions of lifelong learning to adult learners will help them to continue developing on a personal level, having greater individual autonomy and making a more active and productive contributor to society. The role of lifelong learning in the life of adults is clearly visible from a study reported by Brien (2009),

> When a group of older adults, age 55-75, were asked if they would be interested in life-long learning and living in a college atmosphere, more than half of the respondents said they like the idea of retiring to a home on a college campus.

Lifelong learning supports adult learners to remain longer in productive employment and contribute more to work even in their later stage of life. Wolff (2000, p. 10) observes,

> With the decreasing numbers of population aged 20-65, lifelong learning will help the elderly to increasingly remain in the work force, as a means of reducing poverty, increasing economic growth, and giving a stronger sense of self value to the elderly themselves. While the elderly lose some skills, such as working in physically demanding and time-intensive jobs, overall they can continue to engage in occupations ranging from the most rudimentary to the most sophisticated.

Talking further on this issue, Wolff (2000) advocates,

> In many cases, the elderly will be able to use their existing skills. In other cases, they will require training in new skills, such as computer literacy. The elderly also can be trained in new productive roles in areas as varied as childcare, senior adult care, school assistance, security guarding, and conflict resolution(p.10).

Lifelong learning offers numerous choices and opportunities for the elderly as it helps them to overcome social exclusion and isolation, to remain active, to continue their active citizenship, and to utilize their fullest potential for benefit of society. The demand of lifelong learning for welfare of adults is clearly advocated by a publication of AGE (2007),

> We live in an increasingly knowledge-based society with an ageing population and a more intensely competitive global economy. It is therefore vitally important that older people continue to learn, keep up with and adapt to change so as not to be excluded from society.

Talking about the benefits of lifelong learning in the life of adults, Aggett and Neild (2014) suggest that there is a considerable body of evidence that clearly establishes a wide range of benefits that flow from learning—to society, the economy, organisations and individuals. A report from UKCES, *The Value of Skills* concludes that investing in skills and learning benefits to following:

- Society with higher employment, a healthier population, greater civic participation and less crime;
- The economy by increasing the productivity of the workforce and increasing employment rates;
- Organisations by having a more productive and innovative workforce, by being more competitive and more able to adapt to changing economic conditions;
- Individuals by raising their likelihood of being in employment and by leading to improved wages, improved health and well-being and improved resilience to changing economic conditions (UKCES 2010).

Adding further, Aggett and Neild (2014, p. 4) argues that these outcomes are inter-related. According to them, improvements in skill levels can lead to a rise in employment, which reduces poverty. Poverty is linked to illness, disease and unhealthy behaviours; which means raising skills levels lead to reduced public spending on health care. These observations and arguments warrant us to include newer areas to practice lifelong learning and look for finding innovative ways to offer it among adult learners. Mathematics is one such emerging area and the reason is simple. Math is a skill that all adults use every day, whether they realize it or not. Discovering maths later in life can be really important in achieving their potential (U.S. Department of Education 2015).

2.1.2 Lifelong Mathematics Learning for Adult Learners: Perceived Benefits and Challenges

The conception of mathematics implied by adult mathematics education is broad and inclusive, encompassing diverse areas of activity, including: specialized mathematics and service mathematics (as in higher education), school mathematics, vocational mathematics, street mathematics, mathematics for everyday living, and adult numeracy (FitzSimons et al. 2003). Since today's decisions are based on data, it is equally important for adult learners to develop and strengthen skills in mathematics. Mathematics skills are a gatekeeper for further education and training, and significantly affect employability and career options. Even for jobs requiring postsecondary education, employers seek employees who are proficient in mathematics, as well as reading; use math to solve problems; and communicate effectively (Southern Regional Education Board 2015). In addition to economic benefits, mathematics has also been seen as a tool to promote social values and termed as part of our culture. Talking about the benefits of mathematics in social terms, Schlöglmann (2002, pp. 143–144) emphasizes,

> Democratic principles such as equality, justice and so on need an operational concretization. On the one hand, democracy demands a means for communicating and discussing principles in a rational way. Mathematics, with its close relationship to rationality, is our concept to do this. On the other hand, democracy demands operational procedures for its concrete implementation. Mathematics is again the tool that facilitates this.

Emerging economies and technological development in the labour market is the main reason for demanding mathematics education for all including adult learners (FitzSimons 2002). While, Wedege (2010, p. 91) cited a doctoral study of Johansen (2006, p. 275) and observed that Johansen's analysis help us to learn that politicians and educational planners—in their discourses—constructed a common picture of the world with:

- a labour market with demands on adults' [mathematical] knowledge and skills
- an educational system with demands on adults' [mathematical] knowledge and skills

- an everyday life with demands on adults' [mathematical] knowledge and skills
- a societal life with demands on adults' [mathematical] knowledge and skills

(Insertion from Wedege (2010, p. 91).)

Instead of these multi-faceted benefits, adult learners still feel reluctant to lifelong mathematics learning. The reasons are many. First among them is negative perception about mathematics. Many adult learners approach math with anxiety and frustration. Negative previous experiences with math instruction create legitimate barriers for many adult learners (U.S. Department of Education 2015). Mathematics in particular is often associated with negative memories, and so people try to avoid using mathematics in their everyday or vocational lives. This leads to a problematic affective situation in adult-educational mathematics courses (Schlöglmann 2006, p. 15). According to Klinger (2005, p. 7),

> A major challenge for practitioners in adult mathematics education is to achieve effective learning outcomes in the face of prevailing negative attitudes in their students, often present as a consequence of unsatisfactory early mathematics learning experience and flowing from the well established connection between adult innumeracy and mathematics anxiety.

Second, adult learners' everyday competences do not count as mathematics (FitzSimons 2002). Adult learners practice different types of mathematical activities in their everyday life. But learners themselves, employers and societies hardly recognize these activities as mathematical competences. Talking about this tendency of adult learners, Wedge (2010, p. 89) comments, "People simply do not recognize the mathematics in their daily practice—as mathematics. They do not connect the everyday activity and their own competence with mathematics. Most of them only associate mathematics with the school subject". As a result, adult learners do not pay enough attention to improve their mathematical learning by practicing their routine activities.

Third, a major challenge is procedures of mathematic learning surrounded by a popular belief that math is the subject about which students cannot ask "why." In the words of Chisman (2011, p. 7), "The greatest concern of math reform advocates is that most instruction in this field consists of memorizing rote procedures for solving math problems." Too much emphasis on memorizing procedures and too little on conceptual understanding lead to a situation where learners started hating mathematics. The other issue is ability of school teachers teaching mathematics. Teaching mathematics based on rigorous, focused, and coherent standards requires teachers to know mathematics in ways that are likely different from how they were taught. Such teaching requires an understanding of the mathematics taught but also the mathematics that comes before and after that content so that appropriate connections can be established (Dixon 2015). But finding teachers having these types of mathematical abilities is getting more and more difficult.

The above discussions clearly reveal that lifelong mathematics learning is necessity of our times. Promotion of this learning among adult learners offers multiple benefits ranging from personal to social to economical to political. Efforts have been made in different parts of the world to realize this potential but success still eludes us.

The reason is that mathematics education is facing a number of challenges and these are equally applicable to adults learning of mathematics. To know about these challenges, it becomes obvious that one must study different researches about adult mathematics education that are spread across the publications of several disciplines—adult learning, mathematics education, and educational theory—or lies hidden in doctoral dissertations.

2.2 Learning from Research

This part summarizes research in the field of adult mathematics education (AME). It represents the fruit of a literature review that examined doctoral dissertations indexed in *ProQuest Dissertations & Theses Global* published during the period 2000–2015 (100 dissertations), journal articles indexed in the *Proquest Education Journals* (100 articles) under the subject heading "adult" and "mathematics" and "education", and articles published in the Adults Learning Mathematics publications: *ALM International Journal* (www.alm-online.net/publications/alm-journal) and the proceedings for the first 20 ALM conferences (1994 through 2013). The overwhelming majority of the articles were found in the publications of *Adults Learning Mathematics—A Research Forum*. Of the Six themes that emerged from the review, five are pertinent to the troika of adult learners, lifelong learning and mathematics:

1. **Affective Factors—Obstacles to and Advantages of the Adult Learner**: Several studies addressed the challenge of overcoming math and test anxiety and building student self-efficacy to promote success. Motivation and time management skills work in favour of the adult learner.
2. **Theoretical Framework—The Underpinnings of Adult Math Education**: Prominent theorists drew from learning theory, adult theory, and mathematics education theory.
3. **Mathematics for citizenship—Improving in Place**: Under this theme would fall critical pedagogy, parent education and financial literacy. Excluded from this category were studies about workplace and vocational education as these have a separate topic study group at the congress.
4. **Mathematics for Credentialing—Catching Up**: The mathematics taught in elementary and secondary (ages 5–16) is offered at a variety of levels globally. Included here are adult basic and secondary education designated as ABE, ASE, and GED in the US. Developmental mathematics replicates the same mathematical content but in a tertiary institution.
5. **Professional Development—The Teacher as an Adult Learner of Mathematics**: Many studies addressed the education of pre-service and practicing teachers. If we are ever going to break the cycle of poor mathematics learning experiences it starts with confident and knowledgeable school and adult education mathematics teachers.

2.2.1 Affective Factors—Obstacles to and Advantages of the Adult Learner

There is an extensive literature base of research on affective factors in mathematics education although it is not specific to adults. A recent text explored the theories that link beliefs and attitudes about mathematics as well as the emotional and cultural influences on their development (Pepin and Roesken-Winter 2005). Specific to adult education, Schlöglmann discussed the relationship between affective and cognitive aspects of mathematics learning by adults. Citing Ciompi, he situates the learning of mathematics by adults in two realms of research: cognition and psycho-analysis. Schlöglmann states that:

> Adults have many experiences concerning mathematics, especially school mathematics, but most of them have also contact with mathematics in their job and in the everyday life. All these experiences are combined with positive or negative affects and these affects influence their learning processes (Schlöglmann 1999, p. 199).

Evans has also explored the interplay of affect and emotions in his research with adult students. He roots the emotional experiences of students in their cultural experiences and language, particularly their history of involvement in pedagogic practice (Evans 2002).

2.2.1.1 Math Histories

One way to investigate adult mathematics students' earlier mathematics experiences is the use of mathematics histories. These are often used informally by teachers at the beginning of a course as an "ice-breaker" activity to learn something about their students. They have, however, been used formally by several researchers. Thumpston and Coben used semi-structured interviews to explore the math histories of mature students at a London tertiary institution. They found that students often viewed the mathematics they encountered in their work or personal life as being invisible or just "common sense" while they math they could not do was mathematics (Coben and Thumpston 1995 and Coben 1997).

Lindberg used graphs as a tool to gather the math histories of university students who were studying to become mathematics teachers. The graphs and their accompanying narratives identified affective factors that were external to school (life changes), internal to school (interest and motivational changes), external to subject (teaching material and administrative details), and internal to subject (preknowledge, expectations, or the teacher). One observation that she made relates to the teacher, a theme that recurs throughout the AME research: "When the desire to learn mathematics and when the interest in mathematics has been good or excellent the students often have given credit to the teacher (Lindberg 2006, p. 205)."

Whitty used video interviews of her developmental math students to capture their math histories and to solicit input about the characteristics of "good" and "bad" teachers. The pivotal role of the teacher surfaces again in their responses.

The positive impact of initial successes in the course resonated in their increased confidence (Whitty 2010). Whitten interviewed six young adult learners to evaluate their beliefs, attitudes and learning histories. He identified five themes in their responses: Mathematics is a collection of discrete bits of knowledge separate from the real-world, the purpose of studying math was to complete problems correctly, understanding was peripheral to that task, the teacher is the ultimate, external authority, and listening to that authority passively represented learning mathematics (Whitten 2013).

In a rich and detailed doctoral dissertation, Yuen used math histories to uncover the roots of math anxiety in adult students in a developmental class. She too found major themes that emerged from the analysis of the data, themes that echo those found by Whitten: a belief that learning math means following proscribed steps, expectations concerning the role and behaviour of an instructor, roadblocks inherent in the developmental college delivery system, a view of mathematics that is procedural rather than conceptual, importance of rote memorization of skills and prescribed steps, and the need to take control of one's learning.

2.2.1.2 Math Anxiety

It is almost 40 years since Sheila Tobias wrote her landmark book about the phenomenon that has come to be known as "math anxiety." The premise of her research was the conviction that adults were unsuccessful in mathematics classes because of the presence of math anxiety rather than the absence of ability. The cited research on math histories attests to its pervasive existence. In the intervening years, many researchers in adult mathematics education have examined various facets of the phenomenon. Some, including Yuen, have suggested paths to victory for achievement over anxiety.

Yuen suggest five strategies aimed at moving the adult classroom from the teacher-focused structure indicated by students to one that is more learner-centered. Her first suggestion states that students should be given active control in the learning process, Secondly, instruction should have two goals, development of conceptual understanding and then the refinement of procedures useful in similar situations. The fostering of a positive atmosphere and quashing negative, anxiety producing experiences through vigilant, open communication between instructor and students is her third recommendation. Contextualizing problems is her fourth suggestion followed by a fifth and final one, that the instructor helps students see mathematics as a logical activity that can be useful in living an adult life (Yuen 2013).

For her doctoral research, Parker interviewed adults who had overcome math anxiety but had successfully overcome it because the motivation to succeed proved more powerful that the fear of the subject. She found that each individual story followed a pattern, and Parker describes the transition as a six-step journey through and beyond math anxiety:

- Perception of a need to become more comfortable with math
- Making a commitment to address the problem

- Taking specific actions to become more comfortable with math
- Recognizing that a turning point had been reached
- Changing one's the mathematical perspective
- Becoming part of the mathematical support system for other math anxious adults

Parker concluded that overcoming math anxiety during adulthood involves making a transition of major magnitude, that there is an identifiable process, and that a support network is a necessary factor for accomplishing the task (Parker cited in Safford-Ramus 2003, p. 57).

2.2.1.3 Self-efficacy

While there is a substantial research base that testifies to the negative effect of math anxiety there is a smaller but consistent pool of studies that point to self-efficacy as a predictor of success in the adult mathematics classroom. The concept is attributed to the work of Bandura who will be discussed in the section on theoretical frameworks. Stated simply, "Perceived self-efficacy refers to beliefs in one's capabilities to organize and execute the courses of action required to produce given attainments" (Bandura 1997, p. 4). More recently, Dweck speaks of "mindset" and asserts that "The view you adopt for yourself profoundly affects the way you lead your life. A fixed mindset believes that your qualities are static. A growth mindset believes that your basic qualities are things that you can cultivate through your efforts" (Dweck 2006, pp. 6–7). Dweck suggests strategies that promote movement from a fixed mindset to a growth mindset. These include:

- Establish a growth environment.
- Focus on processes.
- Offer constructive criticism that helps the student understand how to fix something.
- Set high standards and help the student reach them.
- For slower students, try to figure out what they do not understand and what learning strategies they do not have.
- Apply the growth mindset to your own teaching. (Dweck 2006, pp. 205–206; cited in Safford-Ramus 2015)

Rowland, in a study of 15 adult undergraduates, found that the following teacher behaviours promoted self-efficacy:

- *Verbal persuasion*, in which the instructor gives a clear statement of his/her philosophy and expectations, continually offers positive reinforcement, and encourages questions at all times.
- *Emotional arousal* is mitigated by a relaxed classroom environment, a patient teacher, content relevant to student lives, and the use of manipulatives.
- *Vicarious learning* was supported by the manner in which course material was presented and by both the teacher and peers modelling successful critical-thinking and problem-solving strategies (Rowland 2004 cited in Safford-Ramus 2015).

Zielke, in a doctoral dissertation, described an intervention designed by him based on successful coaching methods. Using the acronym CHAMP, he devised a program that guided students to move from a state of math anxiety to one of self-efficacy. Using *c*ue words to recall the need to do certain things at certain times, a goal setting focus on the *h*ere and now, *a*rousal control to keep emotions in check in trying situations, *m*odelling and *m*ental imaging of good performance, and praise, persuasion, *p*ositive self-talk his students tackled their math course as if it were a sport event to be won (Zielke 2000).

2.2.2 Theoretical Framework—The Underpinnings of Adult Math Education

Adult mathematics education straddles the borders of many academic disciplines. Benn describes it as "moorland" without clear boundaries, adjoining mathematics, mathematics education, and adult education with education, literacy, philosophy, history, sociology and psychology on the horizon (Wedege et al. 1999). Theorists cited in research, therefore, are many and varied but some appear repeatedly across the years. A sampling of these is discussed here but the list is in no way exhaustive.

2.2.2.1 Adult Learning Theory

Malcolm Knowles is credited with popularizing the term "andragogy" to describe teaching of adults contrasted with "pedagogy" the teaching of children. His model is based on six basic assumptions concerning the divergence of adult learners from children:

- Adults need to know why they need to learn something before undertaking to learn it.
- Adults have a self-concept of being responsible for their own decisions, for their own lives.
- Adults come into an educational activity with both a greater volume and different quality of experience from youth.
- Adults become ready to learn those things they need to know and be able to do in order to cope effectively with their real-life situations.
- Adults are life-centered in their orientation to learning.
- While adults are responsive to some external motivators, the most potent motivators are internal pressures (Knowles et al. 1998, pp. 64–68)

Whether researchers found that their studies agreed or conflicted with Knowles' criteria it still served as the basis of their work. Many of them refer to other theorists like Jack Mezirow, Paulo Freire or Lev Vygotsky and authors of social constructivist theory like Albert Bandura, Jürgen Habermas or Michel Foucault.

2.2.3 Mathematics for Citizenship—Improving in Place

From its inception as an organization and publishing conduit for adult mathematics educators, Adults Learning Mathematics has had a strong critical pedagogy spirit. Many of the founding members had begun their careers as literacy tutors drawn into numeracy at the behest of their students. Others taught at further education or community colleges, institutions that provide a second chance at learning for adults. As a result, mathematics education for empowerment has been implicit in the papers presented or explicit in the annual conference theme. This section of the paper will summarize papers presented on the themes of numeracy for citizenship and, specifically, parenting. Quite by coincidence, the morning newspaper shared a statistic that 69 % of United States parents often struggle helping their children with STEM-related homework (Asbury Park Press, September 8, 2015, p. 1B).

2.2.3.1 Social Issues

Writing of her work with the Landless People's Movement, Gelsa Knijnik stresses the cultural nature of mathematics and the power the subject gives to those who teach and do academic mathematics while subjugating practitioners of indigenous or "street" mathematics (Knijnik 1997). Benn follows a similar path when detailing her journey from a believer that mathematics is value and culture-free to a person questioning the power it holds over the adult population, particularly mature students. In her words, "I became committed to the notion that adult education has a vital role to play in a democratic society. I became convinced that the low level of numeracy in our society limit participation and critical citizenship (Benn 1998, p. 156)." In a separate paper, Benn argues that the education system perpetuates a limiting social class and working-class adults who return to study lack the social support system needed to prevail (Benn 1999). This is reminiscent of Parker's study on successful students—one of the key elements she found was the existence of a backup person willing to shoulder responsibilities so that the adult student is free to study.

Coben's paper on Freire and mathematics education has been cited earlier in this chapter. She has also researched extensively on the Gramscian view of "common sense" and its relation to mathematics education. Like Benn, Coben argues strenuously that knowledge of mathematics is socially powerful and possessing it carries prestige along with an assumption of superior intelligence in general. Common sense, on the other hand, is devalued and individuals who rely on it for mathematical decisions see themselves, and are seen by others, as socially inferior to individuals who can "do" academic mathematics (Coben 1999).

Dias investigated the applicability of Freire's liberation pedagogy to a basic education program in Brazil. The teachers were accustomed to using his work in a literacy program but struggled to transfer that experience to the teaching of mathematics. Teacher discussions revealed "the existence of strong ideological beliefs about how mathematics should be taught, who can learn it, and who knows it

(Dias 2000)." Even teachers committed to critical pedagogy for literacy fell back into traditional views of what mathematics is and how it must be taught.

Yasikawa warns of the external threats to programs that teach mathematics for social justice. Much, if not most, of adult mathematics education takes place in venues funded by governmental or industrial agencies. Shifts in policy can be abrupt, as in the case of a change in the majority party or global economic events. Her paper spotlighted changes in Australian government policy at the time she wrote, but the message is cautionary for anyone committed to teaching adults mathematics for social justice. She warns that "In order to achieve and sustain our social justice commitments, we need to look outwards and work with others to build a larger and stronger network in which adult numeracy and mathematics education plays an important and critical role (Yasikawa 2006, p. 23)."

2.2.3.2 Parents

Adults returning to study mathematics who are parents often state that they want to be able to help their children learn math. Brew (2001) found that the benefits extend beyond assistance with homework to a sense of improving as a role model and altering the math destiny of the next generation. Government agencies, recognizing the opportunity to improve the math skills of children while at the same time those of their parents, have funded parent-child projects.

Civil has worked extensively with parents in Hispanic communities. In an early project she reported leading a series of mathematics workshops for mothers that functioned like a literature club where the women met and discussed informally a topic introduced by Civil. The women developed confidence in themselves as math learners and it flowed over into their home life (Civil 2001). In a later, larger project she worked with parents on topics that were anticipated to be part of their children's classroom experience. The goal was to help the parents understand math better so that they could work with and help their children at home. Reflecting on the courses offered, she states "Providing a safe environment in which their questions and ideas are encouraged and honoured is crucial to their development as *adult learners* of mathematics … as *parents* … and as *advocates* for their children's education (Civil 2002, p. 66)." In a collaboration with Menendez, Civil interviewed parents who had attended a series of math workshops conducted completely in Spanish. They found confirmation of the theories of Knowles, Vygotsky and Freire in their responses and stressed three points in their conclusions: context is important, while they prefer concrete examples adults also want to understand abstract mathematics, powerful affective factors are present even in nonformal instructional settings without pending assessment (Civil and Menendez 2009).

Ginsburg has also focused her research on parents as adult learners of math. In her study, parents and one grandparent from an urban, low-income population worked in tandem with their child on problems drawn from the textbook series in use in the students' class. Ginsburg recommends that teachers of adult learners consider using their students' children's homework as a focal point for their own learning. This leads the parents not only to a better conceptual understanding of mathematics but also allows them to experience the pedagogical changes being implemented in the children's classroom (Ginsburg 2008).

Connolly supervised an Irish program geared to raise the mathematics "comfort level" of parents at schools that serve economically disadvantaged students. These parents typically avoid contact with the school because of their own negative school experiences or the low achievement of their child. During the initial stages of the program the parents worked on math tasks that were a mix of games and calculator activities. Later, the children joined them and the parents functioned as teacher/leaders of the activities. The response to the program from both the parents and children was overwhelmingly positive (Connolly 2009).

Diez-Palomar and Roldan examined a family mathematics program in Catalonia. The foundation of their project was a concept they termed "dialogic space" which they defined as a place built (not literally) by the participants, a place where they felt safe and free to speak and act openly as a person playing a role within the group (Diez-Palomar and Roldan 2010, p. 57). In summarizing their data, the pair expressed a belief that the dialogic space provided an opportunity for adult students to connect their home-based knowledge and academic knowledge. They assert that the parents participated more freely when they felt that they were a member of a group with a particular role to play (Diez-Palomar and Roldan 2010, p. 63).

2.2.4 Mathematics for Credentialing-Catching Up

Adults pursue the study of mathematics at every possible education level. Some attend adult basic education classes because they have little or no formal education or were unsuccessful while in school. Others left secondary school before completing the academic requirements for a diploma or leaving certificate which they now want for personal goals or need for employment. The decision to commence tertiary studies usually dictates enrolment in courses titled "developmental" whose content duplicates the mathematics taught in elementary and secondary school. Finally, degree programs at the tertiary and graduate levels often require a mathematics component. Each of these scenarios is a step forward for the student who brings to the task motivation and life experiences that they may have lacked in their past.

There is extensive overlap in the mathematics content of the instructional settings described above. Entry-level developmental mathematics courses encompass basic arithmetic topics—decimal numbers, operations, ratio and proportion, and measurement. The challenges to the instructor are also similar. The students may have negative memories of school experiences and dread the possibility of failure. Instructional time is condensed—there may be less than forty hours of contact in which to cover the content of eight years of formal school. On a positive note, the students bring a wealth of life experience, intrinsic or extrinsic motivation, and a desire to understand "why", not just "how", the procedures work.

2.2.4.1 Adult Basic and Secondary Education

Perhaps the earliest and most effective United States research in adult basic mathematics education was initiated by teachers in Massachusetts who began in 1992 to sculpt a math curriculum based on the 1989 *Curriculum and Evaluation Standards for School Mathematics*. Their document, *The Massachusetts Adult Basic Education Math Standards* introduced twelve standards and included anecdotes from teachers who had used the standards as well as suggestions for curricular design (Schmitt 1995, p. 33). The project laid the groundwork for later grant-funded curriculum development and a commercial textbook series titled *EMPower*.

Van Groenestijn (1997) has worked in the Netherlands with literate and semi-literate adults. The project, titled *Realistic Mathematics Education* (RME), viewed mathematics as an essential part of adult life and presented math tasks that were drawn from real life situations. In a later paper about the same project, she describes the challenges faced when assessing learning in the ABE system (Van Groenestijn 2001) The RME project is particularly timely as it is sensitive to speakers of other languages, a challenge being faced by countries throughout the EU at present. Haacke (1998) reported on work at the Regional Educative Centre (REC) with independent learning as its focus—students work together on a problem than work independently at their own pace.

In Ireland, O'Rourke suggests guidelines for the development of adult numeracy materials. She lists: building on the learner's prior experience, focusing on context rather than content, strive to develop higher order thinking skills, structure assessments that reflect the knowledge being sought, and emphasize mathematics as a communication vehicle (O'Rourke 1998, pp. 180–181). Colleran devised a program that aimed at building problem-solving skills for a group of unemployed adults. He used action learning for students to explore and solve problems drawn from everyday life and workplace tasks (Colleran 2000).

Hansen, in Denmark, describes the use of a Flex(ible) Ring as a tool for learning a new topic is mathematics by offering a variety of techniques to do so. The center of the ring is a theme from everyday life and the tracks that encircle the theme are various means—videos, worksheets, written assignments—to explore the theme (Hansen 2005). In Germany, Langpaap worked individually with ten female students who he possessed little mathematical knowledge. Each session began with the student describing an everyday life event from her past week that required mathematics. Using that situation as a starting point, he and the student devised a problem and then solved it, introducing math skills as needed to solve the devised problem (Langpaap 2005). Elsewhere in Europe, projects were developed under the European Network for Motivational Mathematics for Adults (EMMA) and the Norwegian Framework for Adult Numeracy.

Hoogland researched the design of a multimedia tool for teaching math in an ABE setting. His recommendations include posing the problem by using photos or film clips, incorporating problem information as text in the photo or possibly a voiceover in the film, posing only questions that would be real or relevant to the student, and building up the "complexity of the situations and not in the complexity of the mathematical concepts (Hoogland 2008. pp. 174–175)." Rumbelow and Nicolaides reported a BBC project that was an online, informal project that aspired to help adults improve their confidence and fluency with numbers in an informal setting. The audience reported using and expressed interest in learning more about eight types of numeracy tasks: calculating discounts, converting currency or weights and measures, helping with homework, splitting restaurant bills and tipping, estimating shopping totals, personal finance, and work-related tasks (2010).

2.2.4.2 Developmental Mathematics

In the United States, adults commencing tertiary study often lack the academic mathematics credentials needed to study collegiate mathematics, courses which are usually required to complete any degree program. Most tertiary institutions offer refresher or "developmental" mathematics courses to raise the student skill level to a point where they can perform at a collegiate level. Because they have never taken the secondary courses or did so years before enrolling at a tertiary institution, most adults place into a developmental mathematics course, often at the most basic level. The situation is not unique to the U.S. Further education colleges in the U.K. and universities in Ireland and Austria have reported interventions that target the under-prepared student. Gill (2011) reported positive results from a one-week intensive review course for mature students at the University of Limerick. A separate venture, a Maths Learning Centre, is described in an article by Gill and O'Donoghue (2011). They detail the rationale for the centre, the multi-pronged resources offered, and the success rate of the students who availed themselves of the facility. Maasz and Schloglmann detailed the situation in Austria tracing their work with adults back to the mid-70s (1996).

Because most community colleges have an open admission policy, they welcome a disproportionate number of the under-prepared population. As a result, they are more likely to need substantial developmental programs, sometimes separate departments within the college. The rate of success is low. Kimura, in a qualitative study, explored the perspective that students and faculty bring to the developmental mathematics classroom. She grouped the results under three headings: Hatred of Math, Magical Thinking and Logical Fallacies; and Doom and Resistance. Of the sixteen students Kimura interviewed, eleven stated that they hated math and shared stories of years of failure that had fueled negative opinions of themselves as learners in general and a lack of self-efficacy. Some of the faculty interviewed recognized this fact and tried to build student confidence and success but admitted that not all colleagues considered this their role. Among the findings that Kimura labelled "Magical Thinking and Logical Fallacies" were the disconnect between student academic skills and the demands of the collegiate classroom, misconstruing the institutional constraints of course requirements, reluctance to seek help and risk being viewed as remedial. Because they did not recognize the course as foundational to success in credit-bearing courses, interviewees set minimal goals—only a "C" grade is needed to pass the course. Finally, what Kimura termed "Doom and Resistance" describes the feeling of abandonment by their K-12 experiences, the institution, and their instructors (Kimura 2012). Rather than being the vestibule to a college degree, for them, and many others like them, developmental math is a moat that the strong can swim but where most college ambitions drown.

Interventions at the developmental level are less likely to be a product of large-scale government projects than adult basic education program research. In fact, most of the research reported at this level lies hidden in doctoral dissertations. A few of those projects were grant-funded but for the most part they appear to be provoked by personal research agendas. Of the 109 dissertations indexed since 2000, 37 were at the developmental level and 10 of those specifically addressed classroom methods.

Students in developmental mathematics courses are disproportionally non-White. Rambish examined a developmental arithmetic course that stressed concept development before procedural competence. She found a significant difference in the course grades for students in the conceptual course versus the tradition-ally taught sections. In particular, Rambish was interested in the performance of African-American students who showed larger gains in the conceptual sections (Rambish 2011). Moreno structured an ethnographic study on the work of Freire, D'Ambosio, and Mezirow and formed a community of learners who learned math-ematics through their personal and shared experiences (Moreno 2011).

2.2.5 Professional Development—The Teacher as Adult Learner

There are two basic categories of teacher as adult learner. The first includes students in undergraduate institutions preparing to become teachers while the second addresses practicing teachers who seek to upgrade their understanding of mathematics and/or best practices for teaching mathematics. Even here there is a blur of borders as the practicing teacher fall into two groups—those who teach children and those who teach adults. The former group sits on the fence between pedagogy and andragogy. All are likely to have similar teacher training as elementary school teachers.

2.2.5.1 Pre-service Teacher Education

Klinger, in two separate journal articles, addresses the weaknesses and needs of this cohort. He presents an impassioned argument for breaking the cycle of innumeracy writing, "If unaddressed, such mathematics aversion will be carried into primary school classrooms, presenting a tangible and substantial risk to the mathematics learning experiences of generations of primary pupils and perpetuating the rela-tionship between adult innumeracy and mathematics anxiety (Klinger 2011, p. 32).

The article proceeds to describe a diagnostic-intervention methodology used in the introductory mathematics course for pre-service teachers. Based on the preliminary diagnostic instrument, focused instruction, sometimes 1-to-1, targeted the conceptual areas in which the student is weak. A reflective exit questionnaire indicated widespread approval by the students.

Recognizing the importance of the critical middle-school years (ages 11–14), Safford-Ramus investigated the state requirements for teaching in grades 6–8 and spearheaded the development and initiation of a trilogy of courses for that pre-service/in-service population. Based on the consensus she derived from the study, the series consists of courses on functions, geometry, and probability and statistics (Safford-Ramus 2010).

2.2.5.2 In-service Teacher Education

During the first decade of this century, policymakers in England supported multiple initiatives to improve adult numeracy, focusing on training efforts for numeracy tutors. Edwards (2010) describes some of the training projects that arose from the initiatives, remnants of which are now housed in the National Institute of Adult Continuing Education (NIACE). Gibney (2010) conducted an action research project with adult numeracy teachers. She devised realistic tasks planned to provoke novel solutions that reflected mathematical thinking.

In the United States, the National Science Foundation funded a numeracy teacher project linked to the *EMPower* series referenced earlier in this chapter and the *Equipped for the Future* project. Adult basic education teachers from six states participated during the five-year life of the project. As in the other international initiatives, the goal of the project was to build teacher confidence through a strong conceptual basis for the procedural mathematics they teach (Schmitt and Bingman 2009). At the same time, The Department of Education Office of Vocational Educational funded the *Adult Numeracy Initiative.* One major product of *ANI* was an environmental scan of the ABE professional development across the country resulting in recommendations for effective PD practices (Safford-Ramus 2007).

This part provided only the briefest synopsis of the work that has been accomplished in the field of adult mathematics education. The intent before presenting all these reviews, however, is to introduce readers to the field and open a door to look at recent developments in adult mathematics/numeracy in terms of policy and provision and discuss some of the paradoxes and tensions that are emerging as adult learning mathematics becomes increasingly regulated in a rapidly developing digital world. How can the research domain of adult learning mathematics develop to be able to connect with the emerging disciplines associated with e.g., technology development. How is numeracy conceptualised and what does this mean for adult learners of mathematics and for their teachers? What kinds of adult mathematics provision are being developed? How is this being translated into practice and what provision is needed for developing teacher knowledge, skills and competence? Chapter 3 discusses all these issues in detail.

2.3 Current Paradoxes, Tensions and Potential Strategies

It would be wrong to say that there is full or uncomplicated consensus when it comes to the issues we grapple with in adult mathematics education. The research domain itself is not clearly defined. The discourse on how numeracy is conceptualised and its relationship with mathematics and literacy is still a matter of debate. There is tension between what policy makers define as numeracy and what is subsequently implemented on the ground through the provision that is offered. How can a community in Ireland, a community in South East Asia or a community in New Zealand conceptualise numeracy and develop associated policy provision to meet the needs of their people? There is clearly no absolute measure, so how do we reconcile the multiplicity of interpretations in policy and provision? This part explores the paradoxes and tensions that exist in the research domain, practice and provision and offers some constructive recommendations to address the issues raised. Thinking about a good definition for the research domain and its boundaries is an important part of working in this area since we have started in the 1990s.

2.3.1 The Disparate and Competing Conceptualisation of Numeracy

There have been many excellent reviews of the conceptualisation of numeracy and its development since the 1990s (see for example Kaye 2010; Coben 2003; Gall 2009). In general terms the conceptualization of numeracy focuses around its relationship with both mathematics and literacy. Maguire and O'Donoghue (2003) developed an organizing framework (Concept Sophistication in Numeracy—an Organising Framework), which considers the development of the concept of numeracy as a continuum with three merging phases: Formative, Mathematical, and Integrative. The phases represent an incrementally-increasing degree of sophistication in conceptualisation. Starting from a very limited concept of numeracy, where it is considered as basic arithmetic skills (formative phase), the framework then moves through to a concept of numeracy as being 'mathematics in context', which recognises the importance of making explicit the significance of mathematics in daily life (Mathematical Phase). The continuum culminates in a conceptualisation which views numeracy as a complex, multifaceted sophisticated construct, incorporating, the mathematics, communication (incl. literacy), and cultural, social, emotional and personal aspects of each individual in context (Integrative Phase).

Coben (2006) rightly points out that although conceptualization of numeracy always includes mathematics is does not work in reverse. Further she highlights how numeracy in some circumstances is conveyed as a component of mathematics e.g., Wedege et al. (1999), and in others, how numeracy is considered to be "not less than maths but more" (Johnston and Tout 1995). Others have highlighted that the acquisition of mathematical skills alone does not constitute numeracy (O'Donoghue 2003, p. 8). Numeracy has been defined as a socially based activity requiring the ability to integrate mathematics and communication skills (Withnall 1995). Other authors have characterized numeracy as a semi-autonomous area at the intersection between literacy and mathematics (Gal 2000, p. 23).

In its earliest conceptualisation numeracy provision was delivered through literacy provision, which has influenced its development as a concept, consequently any discussion of the conceptualisation of numeracy would be deficient without a consideration of this relationship. An evolutionary trail of the concept of numeracy (through a literacy lens), initiated with the Crowther definition of numeracy as 'the mirror image of literacy', has been attempted by O'Donoghue (2002). He sets out the following waypoints in the concepts development:

- Mirror image of literacy.
- Literacy (no explicit concern for numeracy except grassroots interests).
- Literacy (concern for 3R's and basic mathematical skills).
- Functional numeracy (detached from literacy).
- Literacy (numeracy is recognised as an aspect e.g. quantitative literacy).
- Types of literacy (e.g. mathematical literacy, scientific literacy etc.).
- Numeracy detached from literacy equally important (p. 48).

(O'Donoghue 2002)

The relationship between literacy and numeracy is further complicated when one examines the way numeracy is conflated in written texts and other situations involving literacy. Gall (2000) argues that in these situations, literacy can be viewed as a component of numeracy, and Coben (2006) highlights that in these circumstances 'numeracy is more than literacy'. Van Groenestijn (2002) considers numeracy to have its 'own content for every individual person' and to be part of a broader set of knowledge, skills and feelings viewed as a 'backpack' or 'entity' filled with a mix of real life experiences and school knowledge and skills.

Maguire (2003) proposed process model of numerate behaviour. This model is extended here, and provides some insights on the inherent conceptual complexity of this concept. Numeracy is considered as an individual and dynamic attribute. Where numeracy is regarded as an integrated[1] web of interacting[2] elements (communication (including literacy), personal and social development, attitudes, beliefs, values, life experience and motivation) with mathematics at its core. The term 'dynamic' captures the range of interactions that can occur in different contexts, and allows for a shift in the balance of equilibrium of the different elements at different times. Numerate Behaviour results from the internal, dynamic interaction of an individual's mathematics with the other elements of numeracy interacting with a particular context at a given instant in time.

---

[1]Integrate in this context is defined as 'to make into a whole, to complete by adding parts to combine into a whole'.

[2]Interact in this context is defined as 'to act on each other'.

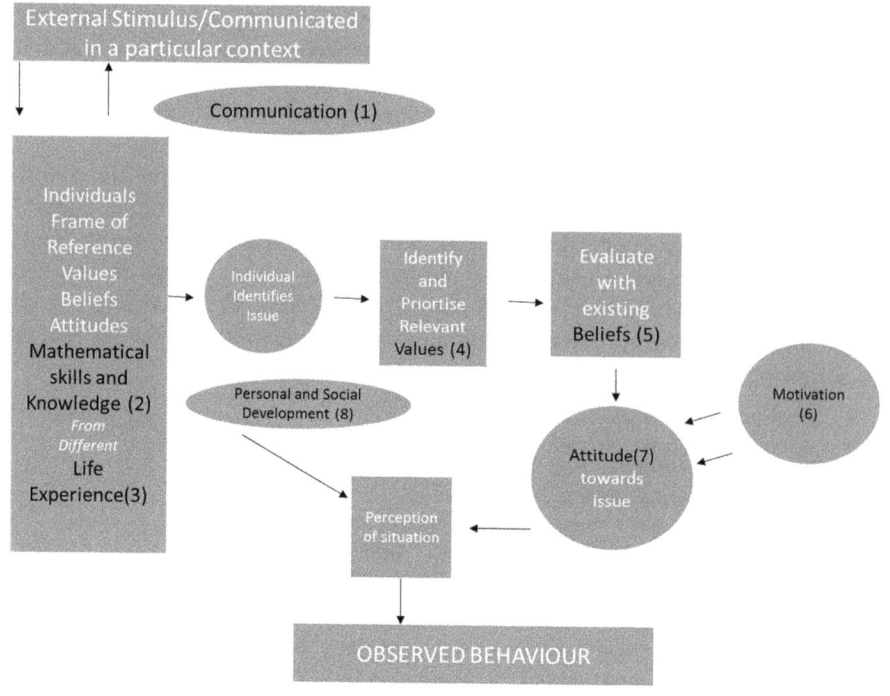

Fig. 2.1 The web of integration and interaction

The process model (Fig. 2.1) put forward for describing how this occurs, has been adapted from a model developed by Peyton (1987) in relation to resource management. It may be described as follows:

2.3.1.1 Communication (1) (Fig. 2.1)

When an individual encounters a numeracy issue in a particular context, he or she perceives the situation from his/her own frame of reference. The individual's frame of reference is a consequence of their life experiences and the consequent values, beliefs and attitudes. In a numeracy situation, the individual's frame of reference in inextricably linked with their mathematical skills and knowledge and their communication (including literacy) skills. At this stage the individual is faced with the task of interpreting the information in whatever form it is communicated which describes the 'issue'. The level of interpretation will be different for each individual and is determined by an individual's facility with the particular content and form of communication and their ability to interpret that information.

2.3.1.2   Identifying the Issue (2) and (3) (Fig. 2.1)

Based on the individual's level of communication and interpretation and directed by the individual's frame of reference, the individual identifies 'the issue'.

2.3.1.3   Value Filter (4) (Fig. 2.1)

Once the issue has been identified the individual identifies and prioritises relevant values. Many adults express feelings, beliefs and values about mathematics that they have developed as a consequence of the mathematics they experienced in school (FitzSimons 1994). At this stage additional information may be sought by the individual to further guide their eventual behaviour. Attitudes change as different values are prioritised or as beliefs are modified by new information or learning (Peyton 1987).

2.3.1.4   Belief Filter (5)–(7) (Fig. 2.1)

The consequences of the value filter are then evaluated against the individual's existing beliefs and directed by their motivation i.e. incorporating both the longer term goals and the individual's instantaneous reaction. These interactions allow the individual to form an attitude towards the issue. Motivation is key to determining the observed behaviour. The term is used to capture the longer-term goals of an individual, their aspirations and ambitions in the context of their own lifelong learning. In the second context the term is used to capture an individual's 'willingness to engage' at a particular instant in time (want to/don't want to, need to/have to). The latter interpretation may cause a particular behavioural outcome that could in fact be running contrary to the long-term goals of the individual concerned.

2.3.1.5   Intervening Variables (8) (Fig. 2.1)

At this stage the outcome of the earlier stages is influenced by intervening variables. In particular, an individual's personal and social development, their skill in making decisions, their confidence, their current priorities, together with their opportunity to act, and the level of support available, will together determine the person's eventual behaviour which will be observable.

The relationship between the web of elements is dynamic and the equilibrium shifts between the elements in different contexts. A positive interaction takes place when the form in which the information is communicated to the individual matches the individual's facility to communicate so that all the relevant information is extrapolated to identify the issue. In a numerate situation the facility to communicate incorporates the ability to interpret mathematical information. A negative interaction will be observed when the (content and form of) communication and the facility to interpret that form of communication are not complementary. Previous negative mathematical experiences may mean that an individual believes that he/she is unable to perform a task and be unwilling or reluctant to attempt it. The situation is even more likely to occur if the task is presented in a way, which is similar to that individual's negative school experience. On the other hand a task that is presented that has relevance to the individual and does not evoke strong negative feeling may be willingly attempted. Synergistic interactions occur more frequently in confident, competent individuals. In this situation, the web of elements has provided a collective of positive interactions that culminate in evoking synergy.

This process model of numerate behaviour clearly has mathematics and literacy as components of numeracy. Building on Coben's (2006) discussion of the relationship between numeracy, literacy and mathematics, the process model for numerate behaviour outlined above and the organising framework it is possible to see the relationship between numeracy, literacy and mathematics in the context of adult mathematics education in an extended way.

In the formative phase, numeracy is regarded a central component of literacy, in the mathematical phase numeracy is a component of both mathematics and literacy. In the Integrative phase mathematics and literacy are central components of numeracy. Numeracy as a concept spans the divide that separates mathematics, real-life, individuals, society and lifelong learning.

However defined, there is a clear need for numeracy to remain a dynamic construct capable of re-conceptualisation according to the contexts in which it is used and by whom. Viewed as a dynamic concept, numeracy can be conceptualised to suit the needs of a community in Ireland, South East Asia or a community in New Zealand and associated policy provision can be developed to meet the identified numeracy needs. However, as outlined the success of policy lies in its implementation. This leads to a third paradox.

2.3.2 Numeracy as an Individual Attribute Versus Legislation for National Curricula and 'One Size Fits All' Policy

The current political and economic environment within which research on adult mathematics is situated is affected by the rhetoric surrounding lifelong learning. Chapter 2 of this publication discusses adults and lifelong learning in detail. However, there is a need to highlight an inherent tension in the current discourse as it has relevance to any attempt to understand the nature of the paradox. A synthesis of the debate regarding the educational meaning of the lifelong learning concept would see it characterised in two ways, each with its supporters. One camp sees lifelong learning as a lifestyle 'ideal', synonymous with terms like: a fulfilled life, personal development and individual liberation. Educationally, this camp see lifelong learning as occurring irrespective of institutions, courses, curricula or teachers. The opposing camp, in the lifelong learning debate, see the concept as having much more formal boundaries. They visualise lifelong learning as a formal and credentialised activity, carried out by recognised institutions and as having measurable outcomes (Burton 2001).

Literacy, numeracy and digital skills are now essential practices that underpin all lifelong learning in a world that is increasingly digital. Together they form the basis for the development of all the necessary qualities for effective participation at work, in the community and in the home (Gall et al. 2009). The results of international surveys including e.g., International Adult Literacy Survey (1997) (IALS), The Adult Literacy and Life Skills Survey (2003–2008) (ALLS) and Programme for the International Assessment of Adult Competencies (2010–2013) (PIAAC) have stimulated a range of policy initiatives to improve adult literacy and numeracy internationally. The implementation of these initiatives is often problematic. It is important at this point to draw a distinction between the term numeracy and the concept of 'numeracy'. Although the term numeracy was not used in every country some conceptualisation of numeracy, can be found in nearly all countries (Maguire 2003). Cultural influences and linguistic differences mean that the specific term numeracy is not always used.

Having a recognised policy for adult numeracy means that numeracy will be tackled head—on in provision. As government agencies address this issue they generally adopt one or other of two approaches. Literacy and numeracy can be integrated into the Vocational Education and Training Sector as part of identified vocational pathways, as workplace learning becomes increasingly important in terms of government priorities (Balatti et al. 2006) or they can be stand-alone 'numeracy curricula'. In the vocational situation, numeracy is often described as a separate element within the standard, but the result usually closely resembles mathematics overlaid with standards and the requirement for measurable learning outcomes rather than not numeracy. Clearly defined numeracy curricula are usually couched in terms of mathematics with an emphasis on context.

Coben (2006) described a way of evaluating numeracy curricula based on a concept she borrowed from Kell (2001). She suggested that the curriculum needs to be considered in relation to the nature of the mathematical demands of adult lives. The purpose of teaching and learning in the wider social, political and economic context and the processes and practices of mathematics learning and teaching with adults. Coben distinguishes between what she describes as Domain 1 and Domain 2 numeracy.

Domain 1 numeracy education is usually associated with a formal standardised curriculum, often with an accredited outcome focus. This domain has what she describes as 'little use value' but is valued by adults and governments and so has 'high exchange value'. In contrast, Domain 2 numeracy education is about mathematics practices and processes in adult' lives which may be informal and non-standard. This domain has what she describes as 'high use value/no exchange value'. Domain Two numerate practices are linked to an individual's 'common sense', their invisible mathematics (Coben 2000) and thus are not always obvious to the individual.

Other alternative views of the numeracy domain have been put forward. Three numeracy lenses were identified by Kanes (2002), *visible numeracy* broadly associated with common mathematical concepts, processes, language which Coben situates in Domain 1 (Coben 2006); usable numeracy is the numeracy of real life problem solving situated in Coben's view in Domain 2. Kanes construct of *constructible numeracy* is "produced by an individual/social constructive process usually in a learning situation" (p. 342) depending on the circumstances could be located within either Domain 1 or Domain 2.

Coben (2006) combined the Domain 1/Domain 2 matrix with the degree of numeracy concept sophistication continuum and suggested that it is possible to position different conceptualisations of numeracy horizontally in terms of concept sophistication and vertically depending on the discursive domain in which they operate. This approach is problematic as it superimposes an operational axis on a conceptual axis. It is not the conceptualisation of numeracy that should be classified but rather its implementation in practice. Revisiting Coben's (2006) approach it is possible to reconceptualise her framework in a more meaningful way.

Domain 1 and Domain 2 might be better considered outputs of informal or formal processes of an individual's numeracy development and view the Domain1 and Domain 2 matrix as a means to evaluate the effectiveness of the implementation of a particular conceptualisation of numeracy in practice. All provision must have the potential from an individual's perspective to have high use/high exchange value (Fig. 2.2). The evaluation must be completed from the perspective of individual adult learners at whom the provision is aimed, as the perceived usefulness and value will be influenced by the starting point and motivation of the adult learner.

In a situation where an adult is returning to adult basic education for the first time a conceptualisation of numeracy as a component of literacy (Formative phase) may well suit their needs. In many countries adult basic education is positioned on a national qualifications framework and could be classified as having high exchange value but with excellent teaching it can also have high use value for the individual learners. Curricula are interpreted and translated into learning opportunity by teachers who play a pivotal role in ensuring high use value to all adult learners. As a dynamic concept numeracy can, and perhaps should, be reconceptualised as adults develop their numerate behaviour during their lifelong learning journey.

Although the subject of initial teacher education for mainstream teachers is well documented in the literature this is much less information on what constitutes excellent adult numeracy teacher education (Morton et al. 2006). FitzSimons (1996) describes the heterogeneity that is characteristic of the teachers in further education in mathematics. She found that there were mathematically highly qualified teachers in practice as well as teachers with low mathematics qualifications,

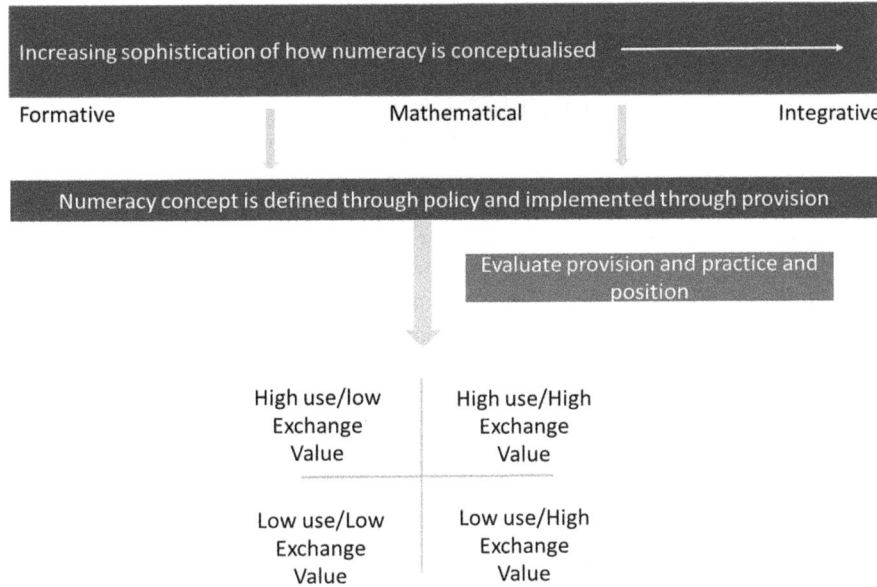

Fig. 2.2 A practical tool for evaluating the implementation of a conceptualisation of numeracy into provision in practice

but who in the first place were well qualified pedagogically. Research in Ireland carried out in 2001 (Maguire 2003) and repeated in partnership with the author and the National Adult Literacy Agency in Ireland in 2013 (NALA 2013) showed very little progress had been made in terms of numeracy teacher qualifications over the ten year period. Numeracy, in the main was taught by part-time teachers with little mathematics or teaching expertise. Other countries e.g. Denmark consider teacher training for adult numeracy education as an add-on to existing teacher qualifications (Lindenskov and Maguire 2005) while others, e.g., England have clearly described development professional pathways for those who teach adult numeracy.

Many Governments reacting to perceived poor results in International surveys are developing policies with resource allocation that become less of a priority over time (Coben 2006). Clearly there are a number of critical factors that influence how numeracy is delivered to adult learners in practice. These are, the policy environment within which teachers must operate, the conceptualisation of numeracy being employed and the appropriateness of the teacher training provided.

The way to ensure that effective adult numeracy provision is available in the context of an individual's own lifelong learning is to ensure that numeracy is conceptualised and reconceptualised to meet the needs of adults at different stages of their lifelong learning journey to ensure high use value/high exchange value. Provision must be developed that captures the vision of this dynamic view of numeracy and teachers must be appropriately trained to be able to deliver numeracy in all its dimensions in their practice to facilitate the development of numerate, self-directed lifelong learners. The practice of numeracy teaching must be considered as being professionally challenging, complex and meritorious in its own right and resourced and developed accordingly.

Applications are contextually bound and that demands teachers who have a wide and deep understanding. There is tension between individualising and defining the solution. The challenge is not to atomise but to take an holistic approach. To maintain relevance to the largest number of learners we need teachers who can manage big ideas. An effective teacher of numeracy has to cross boundaries including e.g., those of mathematics, communication/literacy, technology. This multiplicity of numeracy conceptualisation and provision is inevitable but is it necessarily a bad thing? All too often in looking for solutions the first reaction is to work towards homogenisation as often exemplified in mathematics education, perhaps it is time to accept that heterogeneity is a valid approach to meeting the needs of adult learners. In other words, the challenge before teachers and practitioners in adult mathematics education is to assimilate all these conceptions and approaches to find effective ways to break through the prevailing barriers so that adult learners may experience success in mathematics education on a lifelong basis.

2.4 Promoting Lifelong Mathematics Learning Among Adult Learners: Potential Strategies

Above discussions clearly establishes that efforts should be made at different levels (personal, societal, institutional, and governmental) for promotion of lifelong mathematical learning. Talking about the promotion of lifelong mathematics learning among adult learners, a publication of ALM (Johnson 1998, p. 229) suggests, "It is important that we hold multiple perspectives towards the various goals for adults in learning mathematics and not jump overboard for a particular view of that may simply be a passing fad". Considering this advice, it seems that we have to work on various strategies to promote lifelong learning of Mathematics among adult learners. Five potential strategies that may help to promote lifelong learning of mathematics are discussed one by one.

2.4.1 Promoting Self Directed and Experiential-Learning of Mathematics Among Adults

It is often said that self-learning is best learning and this is more applicable for adult learners. The reason is that adults have various experiences and skills and are supposed to use them for learning new contents, methods or techniques. The key assumptions are that adults tend to be more self-directed, that adults have a vast reservoir of experience that can be used as a resource for learning, that adults are problem centered, and that they need to know why they need to learn something (Merriam et al. 2007, p. 84). While, Frees (2013) observes, "We know that adults learn best through experience, both by taking advantage of the experience they have and by acquiring new experiences; then synthesizing the old with the new to create new meaning." But the question is that how to realize this formula in adult learning of mathematics. Frees (2013) offers a solution, "Two key ways we can engage adult learners, that are active by their nature, are to include opportunities for self-directed and experiential learning."

"Self-directed learning has been seen a process of learning, in which people take the primary initiative for planning, carrying out, and evaluating their own learning experiences" (Merriam et al. 2007, p. 110). Whereas, experiential learning is learning by doing and then reflecting on what was done. These two ways of learning is quite helpful for adult learners, as noted,

> This is just as important as learning the mathematics per se. This should be an integral part of the learning process. That is not to say that adults should not learn skills and techniques, but that they need to understand how such skills and techniques contribute to their personal goals and needs (Johnson 1998, p.229).

Considering the potential, mathematics teachers and researchers are supposed to devise ways and techniques to promote self directed and experiential learning of mathematics among adult learners. Here we must also keep in mind that this type of learning will be a valuable addition to conventional mode of learning i.e. teacher-taught and not a subtraction to it.

2.4.2 Involving Adults (Parents) in Mathematics Education of Their Children

Parental involvement in the form of 'at-home' interest and support has a major influence on pupils' educational outcomes and attitudes (Muir 2002). This is equally true in case of mathematics education. For example, results from a study conducted by Cai (2003) indicated that parental involvement is a statistically significant predictor of their children's mathematical achievement and also promoted positive behaviours and emotional development. He also identified five parental roles in middle school students learning of mathematics: motivator, monitor, resource provider, mathematics content advisor, and mathematics learning counsellor. Similarly a study by Civil (2002) reported that students were of the view that having parents as teachers has proved to be an extremely rich experience and it allows them to learn more about their understandings of mathematics Need of the hour is that we must use all such findings to lure parents to keep learning mathematics to teach their children. This learning will help them in many ways. First, they will be more motivated and committed to understand mathematics and its importance in their lives and strengthen their beliefs about its teaching and learning. Second, as parents themselves learn mathematics with an emphasis on understanding they will become quite vocal about the importance of mathematics education for their children.

Therefore, involving adults (parents) in mathematics education of their children will be a win-win situation. This involvement will help them to understand mathematics from point-of-view of a teacher and learn about its different aspects. Besides, this involvement will also make them more committed and eager to become a lifelong mathematics learner and practitioner.

2.5   Helping Adult Learners to Practice Connectivism in Mathematical Learning

"The challenge for teachers and practitioners in adult mathematics education at any level is to find effective ways to break through the barriers of anxiety and disaffection and to allow students to experience success ……. Ultimately, the greatest achievement will arise when students can overcome their anxiety and aversion to become independent learners with the capacity to extend willingly their engagement with mathematics (Klinger 2010, p. 154)." This requires a paradigm shift in the system of mathematics education. Connectivism, a theory proposed by George Siemens (Siemens 2005) denouncing boundaries of behaviourism, cognitivism, and constructivism holds promise in this direction. This theory works on the principle that a learner who is connected with more content and has more interactivity with the content learns more than anybody else i.e. one who is more connected learns a more than the self. In words of Siemens (2005),

> Connectivism is the integration of principles explored by chaos, network, and complexity and self-organization theories. Learning is a process that occurs within nebulous environments of shifting core elements – not entirely under the control of the individual. Learning (defined as actionable knowledge) can reside outside of ourselves (within an organization or a database), is focused on connecting specialized information sets, and the connections that enable us to learn more are more important than our current state of knowing.

Explaining about the potential usage of connectivism in mathematics education, Klinger (2011, p. 16) observes, "Every new mathematics learning activity should be approached from a language perspective, first identifying a common base of understanding with which students can connect so that concepts can be discussed in natural language before proceeding to translate them into the formalism of symbolic mathematics language." It is supposed that use of connectivism will help math-averse and mathematically anxious adult learners to bring content and content resources together to learn mathematics in a better and understandable way.

2.5.1 Promoting Technology-Based Teaching Learning Activities for Adult Learners

Talking about the use of technology as a lifelong learning tool, European Communities (2010) accept, "Technology not only stimulates creativity and innovation, it also contributes to intercultural dialogue and plays an important role in helping us all overcome our own individual learning challenges." Like different walks of life, technology can be of immense help for promotion of mathematical learning among adults. There are researches that support this claim. For example, Alan (2012) conducted a study to consider the use of mathematical educational software as a means of enhancing the performance rates of adult learners of mathematics in developmental courses and observed that average student in the MyMathLab/MathXL based LE course has an average grade of 2.25 (C) versus the average grade of 1.09 (D) in a traditional LE math course. These results were also substantiated by the claim of all the observed students that they were able to identify the areas that they are deficient in and spend the time on the computer "filling in the gaps" of their learning.

In fact, technological tools offer both content specific and content neutral support in mathematics learning. In mathematics education, content-specific technologies include computer algebra systems; dynamic geometry environments; interactive applets; handheld computation, data collection, and analysis devices; and computer-based applications. These technologies support students in exploring and identifying mathematical concepts and relationships. Content-neutral technologies include communication and collaboration tools and Web-based digital media, and these technologies increase students' access to information, ideas, and interactions that can support and enhance sense making, which is central to the process of taking ownership of knowledge (NCTM 2015). Therefore, more research is necessary to ensure the viability of developmental math learning with the use of technology over learning within the traditional classroom lecture format.

2.5.2 Establishing Lifelong Mathematics Learning Communities for Adult Learners

All over the world different communities, associations and organizations are working to promote adult learning. Parallel to these efforts, we need a new initiative in the form of establishment of Lifelong Mathematics Learning Communities for Adult Learners at local, regional, national and international level. Redecker et al. (2009, p. 11) support this claim,

> Social computing can actively support lifelong learning by offering accessible, flexible and dynamic learning environments that can complement and supplement initial training. Furthermore, the networking potential of social computing, together with its power in overcoming time and space barriers, supports interaction and collaboration among and

between learners and teachers who are geographically dispersed and enables students to broaden their horizons, and collaborate across borders, language barriers, and institutional walls.

These learning communities for elderly can be established online as well as in the form of traditional organizational establishment. The role of media experts and organizations will be to establish and provide technical support to these communities, while adult learners will be required to take care and run these communities. These establishments will act as connecting link for adult learners to fulfil their lifelong mathematical learning needs. These establishments will also provide a forum for the elderly to showcase their expertise and experiences for development of new ways and techniques to practice mathematical learning. In nutshell, these learning communities will be a viable platform for all those adult learners willing to practice mathematical learning on continuing basis or looking forward to share their experiences and expertise to promote it.

Chapter 3
Summary and Looking Ahead

- Significance of troika of adult learners, lifelong learning, and mathematics is described;
- Summarization of the results of a literature review and examination of journal articles indexed as "adult mathematics education" are presented;
- Recent developments in adult mathematics/numeracy in terms of policy and provision and some of the paradoxes and tensions are discussed;
- A number of very useful and pertinent questions regarding adult learning mathematics are raised;
- Five potential strategies for promotion of lifelong learning of mathematics among adult learners are put forward.

© The Author(s) 2016
K. Safford-Ramus et al., *The Troika of Adult Learners, Lifelong Learning, and Mathematics*, ICME-13 Topical Surveys, DOI 10.1007/978-3-319-32808-9_3

References

AGE. (2007). *Lifelong learning-a tool for all ages*. Retrieved from http://www.age-platform.eu/images/stories/EN/AGE_leaflet_lifelong_learning.pdf.

Aggett, K., & Neild, B. (2014). *The benefits of adult learning*. University of Exeter. Retrieved from http://www.learning-southwest.org.uk/wp-content/uploads/2015/03/20141208-Learning-South-West-Soundbites.pdf.

Alan, L. (2011). Technology: The bridge to facilitate learning of adult learners of mathematics. *Proceedings of the eleventh international conference on turning dreams into reality: Transformations and paradigm shifts in mathematics education*. Grahamstown: Rhodes University. Retrieved from http://www.qucosa.de/recherche/frontdoor/?tx_slubopus4frontend%5bid%5d=urn:nbn:de:bsz:14-qucosa-82301.

Balatti, J., Black, S., & Falk, I. (2006). *Reframing adult literacy and numeracy: A social capital perspective*. Adelaide, South Australia: NCVER. Retrieved from www.ncver.edu.au/wps/wcm/connect/b03417bb.../nr4l05.doc?.

Bandura, A. (1997). *Self-efficacy: The exercise of control*. New York: Freeman.

Benn, R. (1997). *Adults count too: Mathematics for empowerment*. Leicester, England and Wales: The National Organisation for Adult Learning NIACE.

Benn, R. (1998). An introduction to adults count too. In D. Coben & J. O'Donoghue (Eds.), *Adults learning mathematics-4* (pp. 156–162). London: Goldsmiths College.

Brew, C. (2001). Implications for women and children when mothers return to study mathematics. In M. J. Schmitt & K. Safford-Ramus (Eds.), *Adults learning mathematics-7: A conversation between researchers and practitioners* (pp. 167–172). Cambridge, MA: NCSALL.

Brien, S. O. (2009). *Baby boomers and seniors embrace lifelong learning campus-based housing offers lifelong learning, a chance to relive college days*. Retrieved from http://seniorliving.about.com/od/housingoptions/a/learning_commun.html.

Burton, L. (2001). Research mathematicians as learners—and what mathematics education can learn from them. *British Educational Research Journal, 27*(5), 589–599.

Cai, J. (2003). Investigating parental roles in students' learning of mathematics from a cross-national perspective. *Mathematics Education Research Journal, 15*(2), 87–106.

Cercone, K. (2008). Characteristics of adult learners with implications for online learning design. *AACE Journal, 16*(2), 137–159.

Chisman, F. P. (2011). *Facing the challenge of numeracy in adult education*. New York: Council for Advancement of Adult Literacy.

CIA. (2015). *The world factbook*. Retrieved from https://www.cia.gov/library/publications/the-world-factbook/fields/2010.html.

Civil, M. (2001). Parents as learners and teachers of mathematics: Towards a two-way dialogue. In M. J. Schmitt & K. Safford-Ramus (Eds.), *Adults learning mathematics-7: A conversation between researchers and practitioners* (pp. 173–177). Cambridge, MA: NCSALL.

© The Author(s) 2016
K. Safford-Ramus et al., *The Troika of Adult Learners, Lifelong Learning, and Mathematics*, ICME-13 Topical Surveys, DOI 10.1007/978-3-319-32808-9

Civil, M. (2002a). Mathematics for parents: Issues of pedagogy and content. In L. Johansen & T. Wedege (Eds.), *Numeracy for empowerment and democracy? Proceedings of ALM-08* (pp. 6–67). Roskilde, Denmark: Roskilde University.

Civil, M. (2002b). Adult learners of mathematics: A look at issues of class and culture. *Proceedings of the 9th international conference of Adults Learning Mathematics* (pp. 13–23). Retrieved from http://www.alm-online.net/images/ALM/proceedings/alm-02-proceedingsalm9.pdf.

Civil, M., & Menendez, J. M. (2009). The role of cognition and affect on adult's participation in a nonformal setting for learning mathematics. In K. Safford-Ramus (Ed.), *A declaration of numeracy: Empowering adults through mathematics education* (pp. 147–164). Lancaster, PA: DEStech Publications.

Coben, D. (1997). Mathematics life histories and common sense. In D. Coben (Ed.), *Adults learning mathematics: Proceedings of ALM-3* (pp. 56–60). London: Goldsmiths College.

Coben, D. (1998). Paulo Freire's legacy for adults learning mathematics. In D. Coben & J. O'Donoghue (Eds.), *Proceedings of adults learning mathematics-4* (pp. 100–108). London: Goldsmiths College.

Coben, D. (1999). Common Sense or good sense? Ethnomathematics and the prospects for a Gramscian politics of adults' mathematics education. In M. van Groenestijn & D. Coben (Eds.), *Mathematics as part of lifelong learning: Proceedings of ALM-05* (pp. 204–209). London: Goldsmiths College.

Coben, D. (2000). Numeracy, mathematics and adult learning practice. In I. Gal (Ed.), *Adult numeracy development: Theory, research, practice* (pp. 33–50). New Jersey: Hampton Press.

Coben, D. (2001). Fact, fiction and moral panic: The changing adult numeracy curriculum in England. In G. E. FitzSimons, J. O'Donoghue, & D. Coben. (Eds.), *Adult and life-long education in mathematics: Papers from Working Group for Action 6, 9th International Congress on Mathematical Education, ICME 9* (pp. 125–153). Melbourne: Language Australia in association with Adults Learning Mathematics—A Research Forum (ALM).

Coben, D. (2003). Adult Numeracy: Review of research and related literature. In J. Maaß & W. Schlöglmann (Eds.), *Learning mathematics to live and work in our world: Proceedings of the 10th international conference on Adults Learning Mathematics* (pp. 78–84). Linz, Austria: ALM & Johannes Kepler Universität Linz.

Coben, D. (2006a). What is specific about research in adult numeracy and mathematics education? *ALM International Journal, 2*(1), 18–32.

Coben, D. (2006b). Surfing the adult numeracy wave: What can we learn from each other in the UK and Australia? In M. Horne & B. Marr (Eds.), *Connecting voices in adult mathematics and numeracy: Practitioners, researchers and learners* (pp. 82–87). Melbourne: ACU/ALM.

Coben, D., & Thumpston, G. (1995). Researching mathematics life histories: A case study. In D. Coben (Ed.), *Mathematics with a human face: Proceedings of ALM-02* (pp. 39–44). London: Goldsmiths College.

Colleran, N., O'Donoghue, J., & Murphy, E. (2000). Improving adult's quantitative problem-solving skills: A new approach. In S. Johnson & D. Coben (Eds.), *Proceedings ofALM-6* (pp. 8–29). Nottingham, UK: University of Nottingham.

Connolly, E. (2009). Fun maths for parents. In K. Safford-Ramus (Ed.), *A declaration of numeracy: Empowering adults through mathematics education* (pp. 32–42). Lancaster, PA: DEStech Publications.

Dias, A. L. B. (2000). Becoming critical of adult numeracy in a Freirian literacy program. In S. Johnson & D. Coben (Eds.), *ALM-6: Proceedings of the sixth international conference of Adults Learning Mathematics—A Research Conference* (pp. 82–91). Nottingham UK: CEP.

Diez-Palomar, J. & Roldan, S. M. (2010). Family mathematics education: Building dialogic spaces for adults learning mathematics. In G. Griffiths & D. Kaye (Eds.), *Numeracy works for life: Proceedings of ALM-16* (pp. 56–66). London: London South Bank University.

Dixon, J. (2015). *Even their mistakes will change.* Retrieved from http://www.nctm.org/Publications/Teaching-Children-Mathematics/Blog/Even-Their-Mistakes-Will-Change/.

Dweck, C. (2006). *Mindset: The new psychology of success.* New York: Random House.

Edwards, M. (2010). From standards-led to market-driven: A critical moment for adult numeracy teacher trainers. *Adults Learning Mathematics—An International Journal, 9*(1), 24–36.

European Commission. (2001). *Making a European area of lifelong learning a reality. Communication from the Commission.* Brussels, Belgium: European Commission. Retrieved from http://aei.pitt.edu/42878/.

European Communities. (2010). *ICT and digital media for key competences EU projects at work.* Retrieved from http://eacea.ec.europa.eu/llp/results_projects/documents/publi/eden_brochure_2010.pdf.

Evans, J. (2002). Developing the ideas of affect and emotion among adult learners. In L.O. Johansen & T. Wedege (Eds.), *Numeracy for empowerment and democracy? Proceedings of ALM-8* (pp.88–96). Roskilde, Denmark: Roskilde University.

FitzSimons, G. E. (2002). *What counts as mathematics? Technologies of power in adult and vocational education.* Dordrecht: Kluwer Academic Publishers.

FitzSimons, G. E., Coben, D., & O'Donoghue, J. (2003). Lifelong mathematics education. In A. J. Bishop, M. A. Clements, C. Keitel, J. Kilpatrick, & F. K. S. Leung (Eds.), *International handbook of mathematics education* (pp. 105–144). Dordrecht, NL: Kluwer Academic Publishers.

Freese, L. (2013). Adult learning in the mathematics classroom: Pursuing discovery learning. *Higher Education.* Retrieved from http://www.pearsoned.com/education-blog/adult-learning-in-the-mathematics-classroom-pursuing-discovery-learning/.

Gal, I. (2000). The numeracy challenge. In I. Gal (Ed.), *Adult numeracy development: Theory, research and practice* (pp. 9–31). Cresskill, NJ: Hampton Press.

Gal, I., Alatorre, S., Close, S., Evans. J., Johansen. L., Maguire, T. et al. (2009). *PIAAC numeracy: A conceptual framework.* OECD Education Working Papers, No. 35, OECD Publishing. Retrieved from http://dx.doi.org/10.1787/220337421165.

Gill, O. (2011). Evaluating the impact of a refresher course in mathematics on adult learners. In H. Christensen, J. Diez-Palomar, Kantner & C. Klinger (Eds.), *Maths at work: Mathematics in a changing world: Proceedings of ALM-17* (pp.37–46). Oslo, Norway: Vox.

Gill, O., & O'Donoghue, J. (2008). Mathematics support for adult learners. In T. Maguire, N. Colleran, O. Gill & J. O'Donoghue (Eds.), *The changing face of adult mathematics education: Learning from the past, planning for the future: Proceedings of ALM-13* (pp. 44–54). Limerick, Ireland: University of Limerick.

Ginsburg, L. (2008). Adult learners go home to their children's homework: What happens when the parent is unsure of the content? In V. Seabright & I. Seabright (Eds.), *Crossing borders—research, reflection & practice in adults learning mathematics: Proceedings of ALM-13* (pp. 55–66). Belfast, NI: Queens University.

Gustafsson, L., & Mouwitz, L. (2004). *Adults and mathematics—a vital subject.* Goteborg: National Center for Mathematics Education.

Hansen, E. P. (2005). Flexible learning. In L. Lindberg (Ed.), *"Bildung" and/or Training: Proceedings ALM-11* (pp. 104–106). Goteborg, Sweden: Goteborg University.

Hoogland, K. (2008). Towards a multimedia tool for numeracy education. In T. Maguire, N. Colleran, O. Gill & J. O'Donoghue (Eds.), *The changing face of adult mathematics education: Learning from the past, planning for the future: Proceedings of ALM-14* (pp. 165–176). Limerick, IE: University of Limerick.

ICME 13. (2015). *TSG 6 Adult learning of mathematics—lifelong learning.* Retrieved from http://www.icme13.org/files/tsg/TSG_6.pdf.

Jarvis, P. (2006). *Towards a comprehensive theory of human learning* (Vol. 1). London, UK: Routledge.

Johansen, L. Ø. (2006). *Hvorforskalvoksnetilbydesundervisning i matematik? En diskursanalytisktilgangtilbegrundelsesproblemet.* [Why offer mathematics education to adults?] (unpublished doctoral thesis). Denmark: Aalborg University.

Johnson, S. (1998). Mathematics for lifelong learning a summary of a plenary discussion around emerging themes at the conference. *Proceedings ALM-5* (pp. 229–231). Retrieved from http://www.alm-online.net/images/ALM/conferences/ALM05/proceedings/ALM05-proceedings-p229-231.pdf.

Johnston, B., & Tout, D. (1995). *Adult numeracy teaching—making meaning in mathematics.* Melbourne: National Staff Development Committee for Vocational Education and Training.

Kanes, C. (2002). Towards numeracy as a cultural historical activity system. In P. Valero & O. Skovsmose (Eds.), *Mathematics education and society, Proceedings of the third international mathematics education and society conference, MES3, 2nd–7th April 2002* (pp. 341–350). Helsingør, Denmark: Roskilde University/Aalborg University.

Kaye, D. (2010). *Defining numeracy, the story continues.* Paper presented at the 17th international conference of adults learning mathematics—A Research Forum (ALM-17). Oslo, Norway.

Kell, C. (2001). Literacy, literacies and ABET in South Africa: On the knife-edge, new cutting edge or thin end of the wedge? In J. Crowther, M. L. Hamilton, & L. Tett (Eds.), *Powerful literacies* (pp. 94–107). Leicester: NIACE.

Kimura, K. (2012). *Doom and resistance: Perspectives of developmental math students at a midwestern community college (Unpublished doctoral dissertation, UMI 3552291).* DeKalb, IL: Northern Illinois University.

Klinger, C. M. (2010). Behaviourism, cognitivism, constructivism, or connectivism? Tackling mathematics anxiety with 'isms' for a digital age. *Numeracy works for life: Proceedings of the 16th international conference of adults learning mathematics* (pp. 154–161). Retrieved from http://www.alm-online.net/images/ALM/proceedings/alm16/Articles/15klinger.pdf.

Klinger, C. M. (2011a). Addressing adult innumeracy via an interventionist approach to mathematics aversion in pre-service primary teachers. *Adults Learning Mathematics: An International Journal, 6*(2), 32–41.

Klinger, C. M. (2011b). 'Connectivism': A new paradigm for the mathematics anxiety challenge? *ALM International journal, 6*(1), 7–19.

Knijnik, G. (1997). Adult numeracy and its relations with academic and popular knowledge. In D. Coben (Ed.), *Proceedings of adults learning mathematics-3* (pp. 30–37). London: Goldsmiths College.

Langpaap, J. (2005). Teaching innumerate adults: Using everyday life experience to develop procedural thinking. In L. Lindberg (Ed.), *"Bildung" and/or Training: Proceedings of ALM-11* (pp. 107–117). Goteborg, Sweden: Goteborg University.

Lindberg, L. (2006). An approach to get to know the mathematical background of the students. In M. Horne & B. Marr (Eds.), *Connecting voices in adult mathematics and numeracy: Practitioners, researchers, and learners: Proceedings ALM-12* (pp. 202–205). Melbourne, Australia: Australian Catholic University.

Lindenskov, L., & Wedege, T. (2001). *Numeracy as an analytical tool in mathematics education and research* (Vol. 31). Roskilde, Denmark: Centre for Research in Learning Mathematics.

Lindenskov, L., Maguire, T., & Weng, P. (2005). *Professional development of adult numeracy tutors—what form does it take? Paper presented at the 10th international conference of adults learning mathematics—A Research Forum (ALM-10).* Austria: Strobel.

Maaß, J., & Schloglmann, W. (1996). Adults learn maths—some results of our research. In D. Coben (Ed.), *Mathematics with a human face: Proceedings of ALM-02* (pp. 26–31). London: Goldsmiths College.

Maasz, J., & Siller, H. (2010). Adults learning mathematics: What we should know about betting and bookkeeping? *ALM International Journal, 5*(1), 52–63.

Maguire, T. M. (2003). Engendering numeracy in adults' mathematics education with a focus on tutors: A grounded approach (Unpublished PhD thesis). Limerick, Ireland: University of Limerick.

Maguire, T., & O'Donoghue, J. (2003). Numeracy concept sophistication—An organizing framework, a useful thinking tool. *Paper presented at the tenth international conference of Adults Learning Mathematics—A Research Forum (ALM-10)*, 29 June–2 July, 2003. Strobl, Austria.

Maguire, T., & O'Donoghue J. (2005). What can school mathematics learn from research on adult learning mathematics? In S. Close, T. Dooley & D. Corcoran (Eds.), *Proceedings of first national conference on research on mathematics education (MEI 1): Mathematics education in Ireland: A research perspective* (pp. 330–346). Dublin, Drumcondra: St Patrick's College.

Medel-Añonuevo, C., Ohsako, T., & Mauch, W. (2001). *Revisiting lifelong learning for the 21st Century.* Hamburg: UNESCO Institute for Education.

Merriam, S. B., & Caffarella, R. S. (1999). *Learning in adulthood*. San Francisco: Jossey-Bass.

Merriam, S., Caffarella, R., & Baumgartner, L. (2006). *Learning in Adulthood: A comprehensive Guide*. San Francisco: Jossey-Bass.

Mezirow, J. (1995). Transformation theory in adult education. In M. R. Welton (Ed.), *In defence of the lifeworld: Critical perspectives on adult learning* (pp. 39–70). Albany, NY: SUNY.

Misra, P.K. (2012). Open educational resources: Lifelong learning for engaged ageing. In A. Okada, T. Connolly, & P. Scott (Eds.), *Collaborative learning 2.0: Open educational resources* (pp. 287–307). USA: IGI Global.

Moreno, G. A. (2011). *A participatory study on an ethnomathematics community in the making (Unpublished doctoral dissertation)*. USA: New Mexico State University.

Morton, T., Maguire, T., & Baynham, M. (2006). *A literature review of research on teacher education in adult literacy, numeracy and ESOL*. London: National Research and Development Centre for Adult Literacy and Numeracy (NRDC).

Muir, T. (2012). Numeracy at home: Involving parents in mathematics education. *International Journal for Mathematics Teaching & Learning*. Retrieved from http://www.cimt.plymouth.ac.uk/journal/muir.pdf.

NALA. (2013). *Doing the maths the training needs of numeracy tutors, 2013 and beyond*. Ireland: NALA.

NCTM. (2015). *Technology in teaching and learning mathematics*. Retrieved from http://www.nctm.org/Standards-and-Positions/Position-Statements/Technology-in-Teaching-and-Learning-Mathematics/.

Niss, M. (1996). Goals of mathematics education in mathematics. In A. J. Bishop, et al. (Eds.), *International handbook of mathematics education* (pp. 11–47). Dordrecht: Kluwer Academic Publishers.

O'Donoghue, J. (2003). Mathematics or numeracy: Does it really matter?. *Adults Learning Maths Newsletter, 18*, 1–8.

O'Rourke, U., & O'Donoghue, J. (1998). Guidelines for the development of adult numeracy materials. In D. Coben & J. O'Donoghue (Eds.), *Proceedings of ALM-4* (pp. 173–191). London: Goldsmiths College.

Pappas, C. (2013). *8 important characteristics of adult learners*. Retrieved from http://elearningindustry.com/8-important-characteristics-of-adult-learners.

Pepin, B., & Roesken-Winter, B. (Eds.). (2015). *From beliefs to dynamic affect systems in mathematics education*. Dordrecht: Springer.

Peyton, R.B. (1987). Mechanisms affecting public acceptance of resource management policies and strategies., *Canadian Journal of Fisheries and Aquatic Sciences, 44*(S2), 306–312.

Rambish, M. C. (2011). *Community college developmental arithmetic course outcomes by instructional delivery approach (Unpublished doctoral dissertation, UMI 3452447)*. Chester, PA, USA: Widener University.

Redecker, C., Ala-Mukta, K., Bacigalupo, M., Ferrari, A., & Punie, Y. (2009). *Learning 2.0: The impact of Web 2.0 innovations on education and training in Europe*. Luxembourg: Office for Official Publications of the European Communities.

Royce, J. (1999). Reading as a basis for using information technology efficiently. In J. Henri & K. Bonanno (Eds.), *Information-literate school community: Best practice* (pp. 145–156). Wagga, Australia: Centre for Information Studies.

Rumbelow, M., & Nicolaides, R. (2010). BBC raw numbers. In G. Griffiths & D. Kaye (Eds.), *Numeracy works for life: Proceeding ALM-16* (pp. 316–325). London: London South Bank University.

Safford-Ramus, K. (2003). A report on selected dissertation research. In J. Maasz & W. Schloeglmann (Eds.), *Learning mathematics to live and work in our world: Proceedings of ALM-10* (pp. 56–69). Linz, Austria: Universitatsverlag Rudolf Trauner.

Safford-Ramus, K. (2007). Professional development in the adult numeracy initiative: A project of the United States of America Department of Education. In V. Seabright & I. Seabright (Eds.), *Crossing borders—research, reflection & practice in adults learning mathematics: Proceedings of ALM-13* (pp. 121–126). Belfast, NI: Queens University.

Safford-Ramus, K. (2008). *Unlatching the gate: Helping adult students learn mathematics.* Philadelphia: Xlibris.

Safford-Ramus, K. (2010). Professional development for middle school teachers: A growing adult student audience. In G. Griffiths & D. Kaye (Eds.), *Numeracy works for life Proceedings of ALM-16* (pp. 224–232). London: London South Bank University.

Safford-Ramus, K. (2015). If self-efficacy deficiency is the disease, what treatments provide hope for a cure? In A. Hector-Mason, Beeli-Zimmerman & G. Griffith (Eds.), *Adults learning maths inside and outside the classroom.*

Schlöglmann, W. (1999). On the relationship between cognitive and affective components of learning mathematics. In M. van Groenestijn & D. Coben (Eds.), *Mathematics as part of lifelong learning: Proceedings of ALM-5* (pp. 198–203). London: Goldsmiths College.

Schlöglmann, W. (2002). Mathematics and society—must all people learn mathematics? In L. Ostergaard & T. Wedege (Eds.), *Numeracy for Empowerment and Democracy? Proceedings of the 8th international conference of adults learning mathematics (ALM8)* (pp. 139–144). Denmark: Roskilde University Printing.

Schlöglmann, W. (2006). A lifelong mathematics learning—A threat or an opportunity? Some remarks on affective conditions in mathematics courses. *ALM International Journal, 2*(1), 6–17.

Schmitt, M. J. (1995). The ABE math standards project: Adapting the NCTM standards to adult education environment. In D. Coben (Ed.), *Mathematics with a human face: Proceedings of ALM-2* (pp. 32–38). London: Goldsmiths College.

Schmitt, M. J., & Bingman, M. B. (2009). Initial findings from research on the TIAN project: A new professional development model for adult education teachers. In K. Safford-Ramus (Ed.), *A declaration of numeracy: Empowering adults through mathematics education: Proceedings of ALM-15* (pp. 190–197). Lancaster, PA: Destech.

Schuller, T., & Watson, D. (2009). *Learning through life inquiry into the future for lifelong learning summary.* Retrieved from http://www.niace.org.uk.

Siemens, G. (2005). Connectivism: A learning theory for the digital age. *International Journal of Instructional Technology & Distance Learning.* Retrieved from http://www.itdl.org/Journal/Jan_05/article01.html.

Southern Regional Education Board. (2015). *Who is the adult learner?* Retrieved from http://www.sreb.org/page/1397/who_is_the_adult_learner.html.

UKCES. (2010). *The value of skills: An evidence review.* UK: UKCES. Retrieved from http://www.ukces.org.uk/assets/ukces/docs/publications/evidence-report-22-the-value-of-skills-an-evidence-review.pdf.

UNESCO. (2015). *Position paper on education post-2015.* Retrieved from http://en.unesco.org/post2015/sites/post2015/files/UNESCO%20Position%20Paper%20ED%202015.pdf.

U.S. Department of Education. (2015). *Mathematics in adult education and literacy.* Retrieved from http://www2.ed.gov/about/offices/list/ovae/pi/AdultEd/math.html.

Van Groenestijn, M. (1998). Constructive numeracy teaching as a gateway to independent living. In D. Coben & J. O'Donoghue (Eds.), *Proceedings of adults learning mathematics-4* (pp. 224–231). London: Goldsmiths College.

Van Groenestijn, M. (2001). Assessment of math skills in ABE: A challenge. In K. Safford-Ramus (Ed.), *A conversation between researchers and practitioners: Proceedings of ALM-7* (pp. 66–71). Cambridge, NA: NCSALL.

Van Groenestijn, M. (2002). *A gateway to numeracy: A study of numeracy in adult basic education.* Utrecht: CD Press.

Vygotsky, L. (1978). *Mind in society.* Cambridge, MA: Harvard University Press.

Wedege, T. (1998). Could there be a specific problematique for research in adult mathematics education? In D. Coben & J. O'Donoghue (Eds.), *Adults learning maths-4: Proceedings of ALM-4, the fourth international conference of adults learning maths—A research forum held at University of Limerick, Ireland, July 4–6 1997* (pp. 210–217). London: Goldsmiths College, University of London, in association with ALM.

Wedege, T. (1999). To know—or not to know—mathematics, that is a question of context. *Educational Studies in Mathematics, 1–3*(39), 205–227.

Wedege, T. (2010). People's mathematics in working life: Why is it invisible? *ALM International Journal, 5*(1), 89–97.

Wedege, T., Benn, R., & Maaß, J. (1999). Adults learning mathematics as a community of practice and research. In M. Van Groenestijn & D. Coben (Eds.), *Mathematics as part of lifelong learning. Proceedings of fifth international conference of adults learning maths—a research forum, ALM-5* (pp. 54–63). London: Goldsmiths College, University of London.

Whitten, D. (2013). Divergent learner pathways: Exploring the mathematical beliefs of young adult learners. In A. Hector-Mason & D. Coben (Eds.), *Synergy: Working together to achieve more than the sum of the parts: Proceedings ALM-19* (pp. 209–219). Waikato, NZ: National Centre of Literacy and Numeracy for Adults.

Whitty, M. (2010). Straight from the student's mouth: Ten freshmen women reflect on their difficulties with mathematics. In G. Griffiths & D. Kaye (Eds.), *Numeracy works for life: Proceedings ALM-16* (pp. 254–264). London: London South Bank University.

Withnall, A. (1995). Towards a Definition of Numeracy. In D. Coben (Ed.), *Adults learning maths—a research forum ALM-1: Proceedings of the inaugural conference of adults learning maths—a research forum* (pp. 11–17). London: Goldsmiths University of London.

Wolff, L. (2000). Lifelong learning for the third age. *TechKnowLogia*, September/October, 9–11. Retrieved from http://www.techknowlogia.org/TKL_Articles/PDF/166.pdf.

Yasikawa, K. (2006). If we have a commitment to social justice, what is it in adult maths/numeracy education that we think is worth fighting for? In M. Horne & B. Marr (Eds.), *Connecting voices in adult mathematics and numeracy: Practitioners, researchers and learners* (pp. 1–4). Melbourne, Australia: Australian Catholic University.

Yuen, C. L. (2013). *Mathematics anxiety learning phenomenon: Adult learners' lived experience and its implications for developmental mathematics instruction* (Unpublished doctoral dissertation). Calgary, AB, Canada: University of Calgary.

Zielke, R. (2000). *The making of a champ: The modification of mathematical self-efficacy beliefs of non-traditional college algebra students using techniques adapted from sports and sports psychology (Unpublished doctoral dissertation, UMI: 9999465)*. Columbus, OH, USA: Ohio State University.

Further Reading

Benn, R. (1997). *Adults Count Too: Mathematics for Empowerment*. Leicester, UK: NIACE.

Coben, D., & O'Donoghue, J. (Eds.). (2009). *Adult mathematics education*. Limerick, Ireland: NCE-MSTL.

Coben, D., O'Donoghue, J., & FitzSimons, G. E. (Eds.). (2000). *Perspectives on adults learning mathematics: Research and practice*. Dordrecht, NL: Kluwer.

Evans, J. (2000). *Adults' mathematical thinking and emotions: A study of numerate practices*. London: Falmer Press.

FitzSimons, G. E. (Ed.). (1997). *Adults returning to study mathematics*. Adelaide, Australia: AAMT.

FitzSimons, G. E., O'Donoghue, J., & Coben, D. (Eds.). (2001). *Adult and lifelong education in mathematics*. Melbourne, Australia: Language Australia.

Griffith, G., & Stone, R. (2013). *teaching numeracy principles and practice*. Maidenhead, UK: Open University Press.

Safford-Ramus, K. (2008). *Unlatching the gate: Helping adult students learn mathematics*. Philadelphia, PA: Xlibris Corporation.

www.ingramcontent.com/pod-product-compliance
Ingram Content Group UK Ltd.
Pitfield, Milton Keynes, MK11 3LW, UK
UKHW020216231225
466357UK00011B/179